AF269773

NUCLEAR
WEAPON BUNKERS
Protecting Stockpiles of Deadly Weapons

by Emily Hudd

CAPSTONE PRESS
a capstone imprint

Bright Ideas is published by Capstone Press, an imprint of Capstone.
1710 Roe Crest Drive
North Mankato, Minnesota 56003
www.capstonepub.com

Library of Congress Cataloging-in-Publication Data
Names: Hudd, Emily, author.
Title: Nuclear weapon bunkers : protecting stockpiles of deadly weapons / Emily Hudd.
Description: North Mankato : Capstone Press, [2020] | Series: High security | Audience: Grades 4-6
Identifiers: LCCN 2019029506 (print) | LCCN 2019029507 (ebook) | ISBN 9781543590609 (hardcover) | ISBN 9781543590616 (ebook)
Subjects: LCSH: Bunkers (Fortification)—Juvenile literature. | Nuclear weapons—Juvenile literature. | Nuclear bomb shelters—Juvenile literature.
Classification: LCC UG405.15 .H83 2020 (print) | LCC UG405.15 (ebook) | DDC 355.7/5—dc23
LC record available at https://lccn.loc.gov/2019029506
LC ebook record available at https://lccn.loc.gov/2019029507

Image Credits
Alamy: Everett Collection Historical, 8–9, Everett Collection Inc, 11, Jon Arnold Images Ltd, 24–25, Robert Burch, 20–21, Trinity Mirror/Mirrorpix, 15; AP Images: Charlie Riedel, 26–27, Juan Carlos Llorca, 5, 28; iStockphoto: ilkercelik, 31; Science Source: Philippe Psaila, 12–13, 16–17, 18–19; Shutterstock Images: gerasimov_foto_174, 6–7, John Wollwerth, cover, 23
Design Elements: Shutterstock Images

Editorial Credits
Editor: Charly Haley; Designer: Laura Graphenteen; Production Specialist: Dan Peluso

All internet sites appearing in back matter were available and accurate when this book was sent to press.

Printed in the United States of America.
PA99

TABLE OF CONTENTS

CHAPTER 1

SAFE AND Secure

Nuclear weapon bunkers are secure. They have strong doors. Some have two fences. Guards stand outside.

A bunker is a place for storing things.
Bunkers keep things safe. Nuclear weapon
bunkers are underground. They have several
rooms. Tunnels connect them.

The nuclear weapons inside the bunkers
are bombs. Bombs cause a lot of damage.
A **missile** is a weapon with a bomb on it.
It carries the bomb.

A soldier walks toward an old nuclear weapon bunker.

a nuclear weapon

These weapons are dangerous. The government must guard them. It keeps facts about bunkers secret. This helps keep the weapons safe.

Workers check a
nuclear weapon.

BUNKER WORKERS

Only workers go inside the bunkers. Cameras record the area. They track people who come and go.

The workers guard the weapons. Some fix the weapons.

HISTORY OF Nuclear Weapons

People made nuclear weapons during the Cold War (1945–1991). There were 70,300 in 1986. Several countries had these weapons. People were scared. The weapons could hurt or kill people.

During the Cold War, some people built emergency shelters to protect themselves just in case a country used nuclear weapons.
Kidde Kokoon
CAN
CANNED WATER

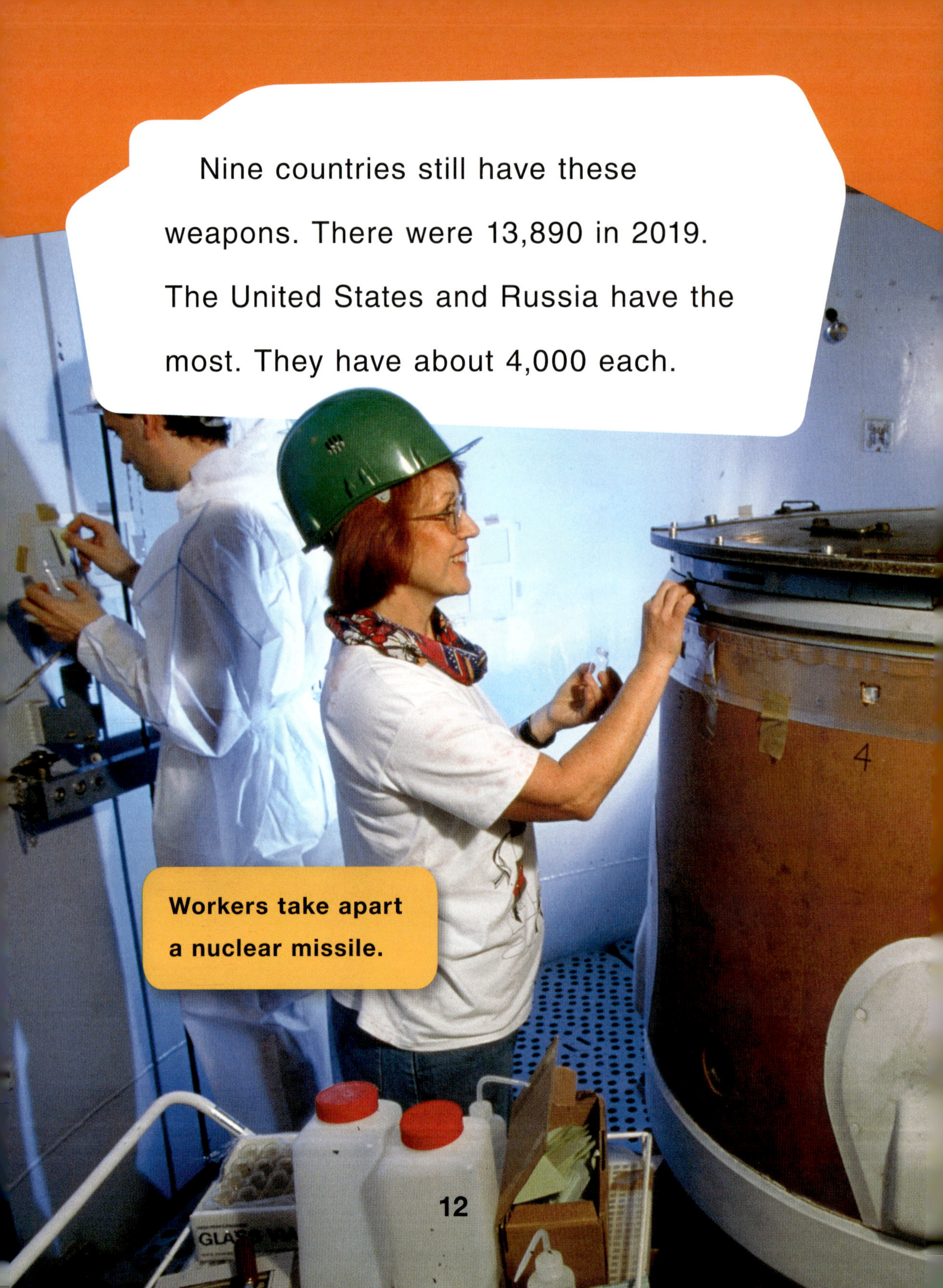

Nine countries still have these weapons. There were 13,890 in 2019. The United States and Russia have the most. They have about 4,000 each.

Some of the weapons are ready to use. Others will be taken apart. Many are being stored.

NORTH KOREA

North Korea has nuclear weapons. Other countries do not know how many it has. The country keeps this number secret.

CHAPTER 3

UNDERGROUND Bunkers

People can only visit old bunkers. Those have no weapons now. One in North Dakota is a museum.

The museum has metal desks and
computers. People worked there years
ago. The computers controlled weapons.

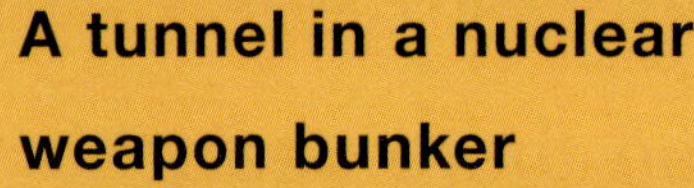

WHERE ARE THE BUNKERS?

Nuclear weapons are in 10 U.S. states. Most are underground. The bunkers are at military bases. The weapons are safe. Workers can fix them. They can take them apart. They make sure the weapons do not explode.

FENWICK

Keeping the weapons in different states protects them. One place could be attacked. But weapons in other places would be safe.

MOVING NUCLEAR WEAPONS

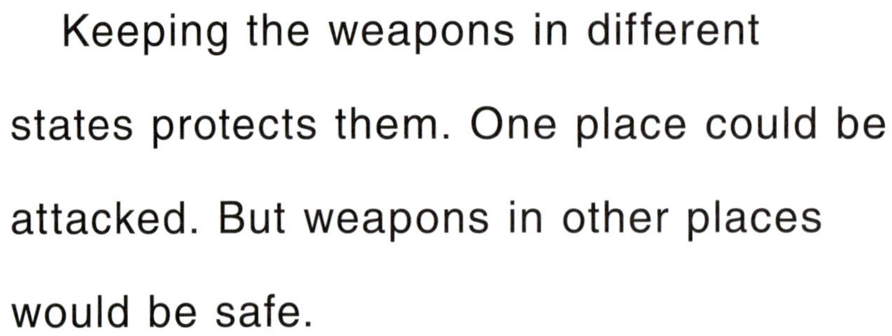

Government workers move weapons around in big trucks. They drive many miles each year.

Workers help move nuclear
weapons to different places.

A nuclear strike bomber aircraft at Kirtland Air Force Base

Kirtland Air Force Base is in New Mexico. It has nuclear weapons. It has the biggest U.S. **stockpile**. But the number there is secret.

SILOS

Many bunkers have **silos**. Silos hold missiles underground. They are connected to other parts of the bunker.

A nuclear missile
in a silo

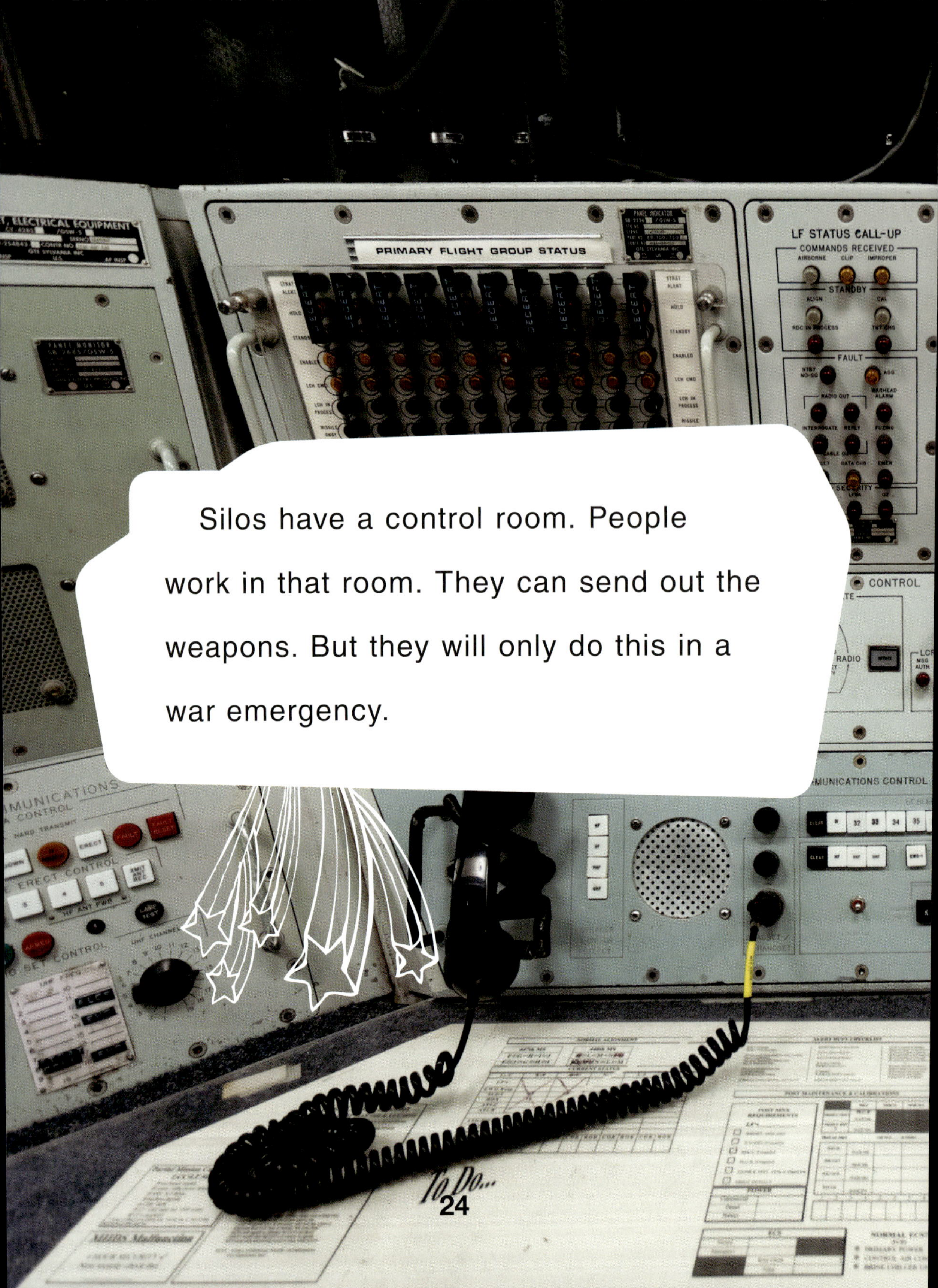

Silos have a control room. People work in that room. They can send out the weapons. But they will only do this in a war emergency.

An old nuclear weapon bunker control room

Five countries in Europe store U.S. nuclear weapons. The places are safe. They have cameras and thick walls. One place has 300 guards.

Nuclear weapons are very dangerous. That is why they must be kept safe.

NUCLEAR WEAPONS UNDERWATER

Some weapons are carried on ships called submarines. These ships go underwater.

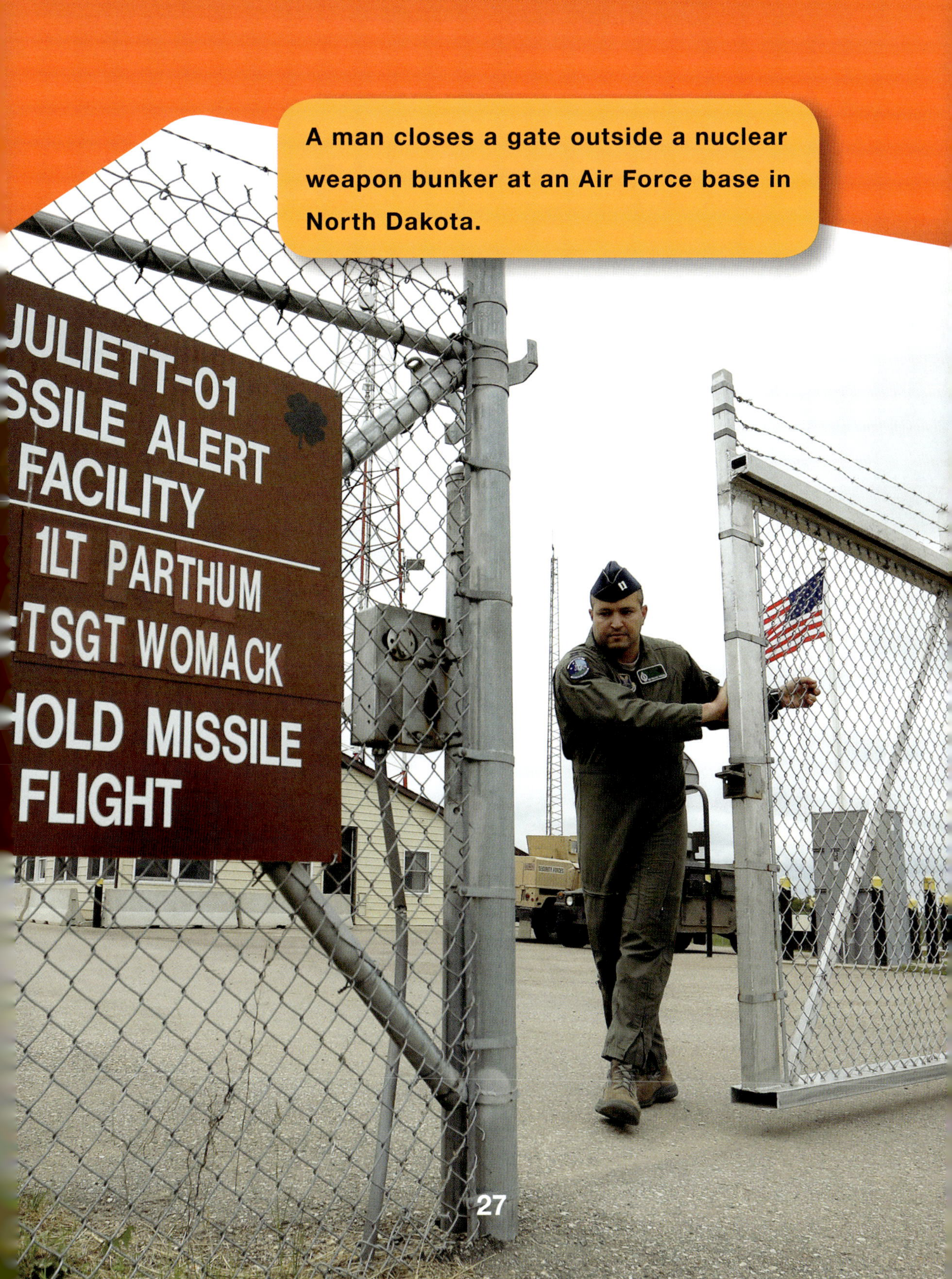

A man closes a gate outside a nuclear weapon bunker at an Air Force base in North Dakota.

GLOSSARY 

missile
a weapon that carries and guides bombs

nuclear weapon
a type of very powerful bomb

silo
an underground storage place for missiles

stockpile
a collection of things in storage

1. Belgium is one country in Europe that stores U.S. nuclear weapons. It has up to 20 bombs.

2. Two U.S. naval bases in Georgia and Washington have nuclear weapons on submarines. The submarines are also called boomers.

3. The last time nuclear weapons were used was in 1945. The United States used the bombs against Japan during World War II (1939–1945). Many people died.

ACTIVITY

WHERE WOULD YOU PUT A BUNKER?

Nuclear weapons are stored in different spots around the country. This helps keep them safe. If you had something important to protect, where would you put it? Imagine you have many things you want to protect. Where would you hide them? Maybe you would hide some underground. Others could be in mountains or in a forest. What other hiding places can you think of? Draw a map of where you would hide your items.

Interested in learning more about nuclear weapons? Look at these resources:

Benoit, Peter. *The Nuclear Age*. New York: Children's Press, 2012.

BrainPOP: Cold War
https://www.brainpop.com/socialstudies/worldhistory/coldwar

DK Findout! The Atomic Bomb
https://www.dkfindout.com/us/history/world-war-ii/atomic-bomb

Want to learn more about where nuclear weapons are stored? Check out this information:

Russo, Kristin J. *Surprising Facts about Being an Air Force Airman*. North Mankato, Minn.: Capstone Press, 2018.

Wonderopolis: How Does a Submarine Work?
https://www.wonderopolis.org/wonder/how-does-a-submarine-work-2